Ecosystems
Energy Flow and Use

by Kate Boehm Jerome

Table of Contents

Millmark
EDUCATION

The Amazon rainforest provides a home for many different kinds of animals. The animals get what they need to live from the rainforest.

Discuss the rainforest pictures with questions like these:

What is the jaguar doing?

The jaguar _____.

Where is the Amazon River?

It's _____.

Why is the tarantula eating?

It's eating because _____.

Share what you know about other animals that live in the rainforest.

jaguar

tarantula

capuchin monkey

The Amazon River in the rainforest in Brazil

kinkajou

Surviving in the Rainforest

Earth has many amazing creatures! One interesting animal is a kinkajou. It lives in the Central American Rainforest.

The kinkajou gets food and **shelter** from the rainforest. It lives in trees and eats fruit. It looks for food at night.

The rainforest is an **ecosystem**. The rainforest ecosystem is warm and has a lot of rain. Since the rainforest is warm and wet, many plants and animals can live and grow there.

shelter – a place to live

ecosystem – an area with living and nonliving things

Living and Nonliving Things

Plants and animals are living things. Most living things need other living things to **survive**. For example, the sloth needs trees for shelter. The poison dart frog needs insects for food.

Living things also need nonliving things to survive. Nonliving things include air, water, sunlight, and soil. Plants and animals need these nonliving things.

The rainforest ecosystem has it all! It provides the living and nonliving things that plants and animals need to survive.

survive – continue to live

sloth

poison dart frog

MEXICO
BELIZE
Belmopan
GUATEMALA HONDURAS
Guatemala City Tegucigalpa
San Salvador
EL SALVADOR NICARAGUA
Managua

ATLANTIC OCEAN

PACIFIC OCEAN

San José
COSTA RICA PANAMA
Panama City

Central American Rainforest

N W E S

0   150   300 Miles
0   150   300 Kilometers

SHARE IDEAS What does the rainforest ecosystem provide for animals? **Explain**.

Populations

There are many different kinds of living things in the rainforest. Each kind of living thing is part of a group, or **population**, in the rainforest.

For example, many different kinds of ants live in the rainforest. One kind of ant is the leaf cutter ant. All the leaf cutter ants in the rainforest form a population. Other kinds of ants in the rainforest form different populations.

The rainforest ecosystem includes many other populations, too. In fact, scientists are still finding new populations of plants and animals in the rainforest.

▲ All the leaf cutter ants in an area form a population.

population – a group of one kind of living thing in the same area

By The Way...

Leaf cutter ants are strong! They can carry leaves that are twenty times heavier than their own bodies.

Communities

All the populations living in a rainforest form the rainforest **community**. This means that different bird populations are part of the rainforest community. Different plant populations are also part of the community. The community is the living part of the rainforest ecosystem.

community – all the populations in an ecosystem

▼ **Many populations of plants and animals live in the rainforest.**

macaws

iguana

banana tree

Interactions

In the rainforest community, different populations **interact** with each other. For example, hummingbirds eat certain flower populations. Tree populations provide shelter for many animal populations.

Scientists study how populations interact. This helps them decide if an ecosystem is healthy.

Trees provide shelter for birds.

---

interact – to act on each other

hummingbird

> **KEY IDEA** In a community, plant populations and animal populations interact with each other.

OBSERVE

Take a walk around your school. Look for plant and animal populations. In a notebook, draw a picture of each population that you see. If possible, write the name of the population under your picture.

Share your notebook with a friend. Ask and answer these questions with each other.

1. What plant populations did you observe?

2. What animal populations did you observe?

3. Name other places where you could find these populations.

MAKE CONNECTIONS

Think of different places you know. What place has the most plants and animals? Draw a picture of that place. Label the plant and animal populations.

USE THE LANGUAGE OF SCIENCE

What is a rainforest community?

A rainforest community is the living part of a rainforest ecosystem. It includes all the populations in the area.

CHAPTER 2

Getting Energy in the Prairie

You can often see a long way in a prairie ecosystem. That's because this kind of ecosystem includes many grasses and bushes. There are no forests with tall trees to block the view!

Like most plants, the grasses and flowers in the prairie ecosystem don't need to eat. That's because most plants use **energy** from the sun to make their own food.

Most plants are **producers**. The food they produce, or make, provides the energy they need to live.

energy – the ability to do work

producers – living things that make their own food

Animals and many other living things are **consumers**. They get energy from the food they consume, or eat. They cannot make their own food.

Some consumers eat producers. For example, a rabbit eats grass. Some consumers eat other consumers. For example, a fox eats mice.

Consumers are grouped according to what they eat. They include herbivores, carnivores, and omnivores.

Explore Language

Latin Word Roots
omnivore
omnis (all) + *vorare*
(to devour, or eat) =
one that eats all

consumers – living things that eat other living things to stay alive

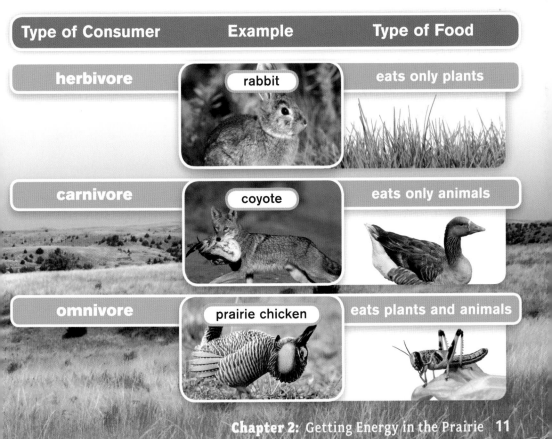

Type of Consumer	Example	Type of Food
herbivore	rabbit	eats only plants
carnivore	coyote	eats only animals
omnivore	prairie chicken	eats plants and animals

Food Chains

Why does a snake eat a mouse? It's because the snake needs food energy from the mouse to live.

In every ecosystem, energy moves from one living thing to another living thing through a **food chain**.

One prairie ecosystem food chain is shown below. The sun supplies energy for the prairie sage to make its own food. Then a grasshopper eats the prairie sage. Next, a prairie dog eats the grasshopper. Finally, an eagle eats the prairie dog. Energy from the sun moves all the way through the food chain.

food chain – the path that energy takes when moving from one living thing to another living thing

A Prairie Food Chain

sunlight

prairie sage

grasshopper

SHARE IDEAS Describe what happens to energy in each step of this food chain.

Decomposers

Another group of living things is important in every food chain. They are the **decomposers**. Decomposers include things such as mushrooms, worms, and **bacteria**.

When living things die, they are decomposed, or broken down, by the decomposers. This returns important **nutrients** to the soil. Plants can use these nutrients when they make food. The food chain begins again.

▼ **Mushrooms that grow on dead trees are decomposers.**

decomposers – living things that break down dead plants and animals

bacteria – tiny living things

nutrients – things that help plants and animals grow

prairie dog

eagle

Food Webs

There are many food chains in an ecosystem. Producers and consumers can be part of more than one food chain. A diagram called a **food web** shows how some food chains connect.

The arrows in the diagram show how different food chains connect. For example, the eagle gets energy by eating squirrels, snakes, rabbits, and prairie dogs. These animals are in different food chains, but the food web shows that they are also connected.

food web – a diagram that shows how different food chains connect in an ecosystem

A Prairie Food Web

eagle

coyote

snake

rabbit

grasshopper

prairie dog

squirrel

prairie sage

KEY IDEA Food chains and food webs show how energy moves through an ecosystem.

COMMUNICATE

Look at the food chain on pages 12–13. With a friend, take turns asking and answering these questions.

1. How does the eagle get the energy it needs?

The eagle gets energy from _____.

2. If all the eagles die, do you think there will be more prairie dogs or fewer prairie dogs? Explain.

I think there will be _____ because _____.

3. If all the eagles die, do you think there will be more grasshoppers or fewer grasshoppers? Explain.

I think there will be _____ because _____.

MAKE CONNECTIONS

Most animals in a food chain are consumers. You are a consumer, too. Do you eat rice? Rice is a plant, so it is a producer. Make a list of other producers that you eat.

STRATEGY FOCUS

Determine Importance

Look at pages 10–14 again.
What are the most important ideas in this chapter?
How do you know?

CHAPTER 3

Energy Use in a Community

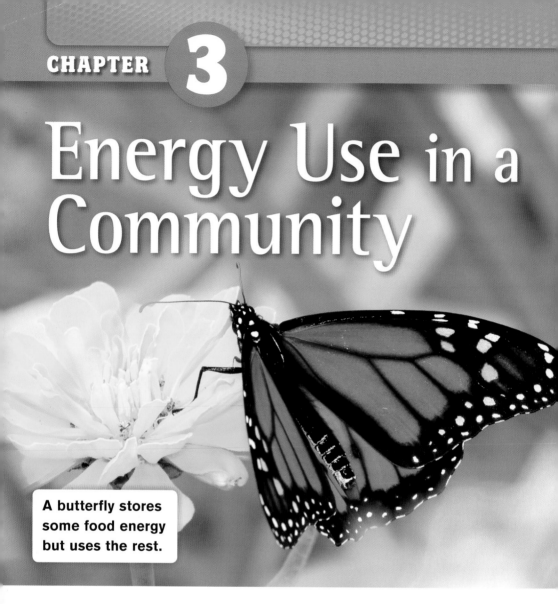

A butterfly stores some food energy but uses the rest.

In a food chain, energy moves from one thing to another. For example, when a butterfly eats nectar from a flower, it gets food energy. The butterfly stores some of this energy in its body.

But some energy leaves the food chain. For example, the butterfly uses energy to fly, to breathe, and to eat. This energy leaves the food chain.

◀ The mouse will not get all the energy that was available to the butterfly.

When a mouse eats the butterfly, the mouse gets food energy. But the butterfly already used some of its energy. This means less energy is **available** to the mouse.

The mouse stores some food energy. But the mouse also uses energy to live. Some energy is lost as heat and waste matter.

▼ The hawk will not get all the energy that was available to the mouse.

When a hawk eats the mouse, it gets some food energy. But it does not get all the energy that was available to the mouse. That's because energy is lost at each level of the food chain.

available – able to be used

SHARE IDEAS Why is energy lost at each level of a food chain? **Explain**.

Energy Pyramid

An **energy pyramid** shows that energy is lost at each level of the food chain. This means there is more energy available at the bottom level, or producer level, than at the higher consumer levels. In fact, scientists think only about one tenth of the energy available at one level passes to the next level!

energy pyramid – a diagram that shows that less energy is available as you move through different levels of the food chain

A Forest Energy Pyramid

CONSUMER LEVELS

PRODUCER LEVEL

owl

mice

crickets

grass

▼ At each level of a food chain, some energy is lost.

KEY IDEA An energy pyramid shows that the bottom levels of a food chain have more energy available than the top levels.

PREDICT

A fire changes a forest ecosystem. Look at the energy pyramid on page 18. With a friend, ask and answer the following questions. Make predictions about how a fire would change a forest.

1. What would a fire do to the producers in a forest ecosystem?

A fire would _____ .

2. After a fire, would there be more energy or less energy available in a forest food chain? Make a prediction and explain your answer.

I predict there would be _____ energy after a fire because _____ .

MAKE CONNECTIONS

Think about the area where you live. Name some producers in your ecosystem. Explain how a fire or a flood could hurt the producers in your ecosystem.

EXPAND VOCABULARY

An energy pyramid is an important concept in science. You may also study other types of pyramids. Find pictures of these examples of pyramids:

1. a geometric figure

2. a food pyramid

3. a large monument from ancient times

Describe each type of pyramid. Explain how these pyramids are alike and different.

Bird Rescue!

You are part of an ecosystem. You can help an ecosystem, too. Some people help ecosystems by protecting the animals in a food chain.

For example, some people work in rescue centers for **birds of prey.** Birds of prey are birds that hunt other animals. They include owls, hawks, and eagles.

Birds of prey are important animals in a food chain. Since they hunt, they keep other animal populations from growing too large. But sometimes these birds are hurt. They need help to survive.

At a rescue center, people take care of injured birds of prey. Some are doctors. They have many years of special training. Other people have no special training. But everyone wants to help. They know that these birds are an important part of the ecosystem.

• Would you like to work at a rescue center for birds of prey? Tell why or why not.

birds of prey – birds that hunt other animals

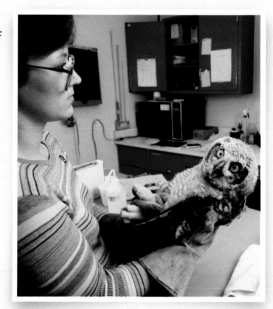

▶ **A rescue center worker helps an owl.**

Words that Explain

When you communicate, you may need to explain the **cause**, or the reason why something happens. You may also need to show the **effect**, or what happens. Words such as **because**, **since**, and **so** help explain causes and effects.

Use **because** and **since** to explain the **cause**, or why something happens.

EXAMPLES The monkey eats **because** it needs energy.

    effect   cause

Since plants make their own food, they don't need to eat.

            cause   effect

Use **so** to explain the **effect**, or what happens because of something else.

EXAMPLES Rabbits eat only grass, **so** they are herbivores.

        cause   effect

Write an Explanation

Choose one ecosystem. Find out about some of the carnivores in that ecosystem.

- Describe the ecosystem.
- Describe some carnivores in that ecosystem. Tell what they eat.
- Explain the role of the carnivores in that ecosystem. Tell why the carnivores are important.

Words You Can Use	
Nouns	**Cause and Effect Words**
carnivore	because
diet	since
food chain	so
food web	therefore
population	as a result
community	

What's in a Taco?

Think about what is in a beef taco. Let's trace each main ingredient back to its producer.

taco

beef

1. Beef comes from cows.
2. Cows eat grass.
3. Grass is a producer.

corn tortilla

1. A corn tortilla is made from corn flour.
2. Corn flour is made from corn.
3. Corn is a producer.

Read the poster.

- Think of your favorite food.
  Try to trace it back to its producers.

Key Words

bacteria tiny living things
Bacteria help break down dead plants.

community (communities) all the populations in an ecosystem
A prairie **community** includes grasshoppers.

consumer (consumers) a living thing that eats producers or other consumers
A cricket is a **consumer** that eats grass.

decomposer (decomposers) a living thing that breaks down dead plants and animals
A mushroom on a dead log is a **decomposer**.

ecosystem (ecosystems) all the living and nonliving things in an area
A rainforest **ecosystem** has many plants and animals.

energy the ability to do work
A butterfly needs **energy** to live.

energy pyramid (energy pyramids) a diagram that shows that less energy is available as you move through the food chain
Grass is at the bottom level of a forest **energy pyramid**.

food chain (food chains) the path that energy takes when moving from one living thing to another living thing
Grass is at the bottom of a forest **food chain**.

food web (food webs) a diagram that shows how different food chains connect in an ecosystem
A **food web** shows how many food chains connect.

population (populations) a group of one kind of living thing in the same area
The prairie ecosystem includes a **population** of eagles.

producer (producers) a living thing that makes its own food
Most plants are **producers** that make their own food.

Index

MILLMARK EDUCATION CORPORATION
Ericka Markman, President and CEO; Karen Peratt, VP, Editorial Director; Lisa Bingen, VP, Marketing; David Willette, VP, Sales; Rachel L. Moir, VP, Operations and Production; Shelby Alinsky, Editor; Mary Ann Mortellaro, Science Editor; Amy Sarver, Series Editor; Betsy Carpenter, Editor; Guadalupe Lopez, Writer; Kris Hanneman and Pictures Unlimited, Photo Research

PROGRAM AUTHORS
Mary Hawley; Program Author, Instructional Design
Kate Boehm Jerome; Program Author, Science

BOOK DESIGN Steve Curtis Design

CONTENT REVIEWER
Carla C. Johnson, EdD, University of Toledo, Toledo, OH

PROGRAM ADVISORS
Scott K. Baker, PhD, Pacific Institutes for Research, Eugene, OR
Carla C. Johnson, EdD, University of Toledo, Toledo, OH
Margit McGuire, PhD, Seattle University, Seattle, WA
Donna Ogle, EdD, National-Louis University, Chicago, IL
Betty Ansin Smallwood, PhD, Center for Applied Linguistics, Washington, DC
Gail Thompson, PhD, Claremont Graduate University, Claremont, CA
Emma Violand-Sánchez, EdD, Arlington Public Schools, Arlington, VA (retired)

TECHNOLOGY
Arleen Nakama, Project Manager
Audio CDs: Heartworks International, Inc.
CD-ROMs: Cannery Agency
ResourceLinks Website: Six Red Marbles

PHOTO CREDITS Cover and 18b © Arco Images/Alamy; IFC and 15b © David Safanda/iStockphoto.com; 1 © Johnny Lye/Shutterstock; 2–3 © Jacques Jangoux/Alamy; 2 and 17a © W. Perry

Conway/Corbis; 3a © Francesco Tomasinelli/Natural Visions; 3b and 5c by Mapping Specialists; 3c © Amazon-Images/Alamy; 4–5 © Peter Oxford/Naturepl.com; 5a © BRUCE COLEMAN INC./Alamy; 5b © Roy Toft; 6a © Bryan Mullennix; 6b © Tim Flach; 7a © Science Faction; 7b © Photofrenetic/Alamy; 7c © Sue Cunningham/DanitaDelimont.com; 8a © Steve Winter; 8b © Michael & Patricia Fogden/Corbis; 8c © Royal Geographical Society/Alamy; 9a © Ellen McKnight/Alamy; 9b and 9c Lloyd Wolf for Millmark Education; 10–11 © Mark Romesser/Alamy; 11a and 14e © Jerry Mercier; 11b © fotosav (Victor & Katya)/Shutterstock; 11c © Eda Rogers; 11d © Elena Kalistratova/Shutterstock; 11e © Tom and Pat Leeson; 11f © Digital Zoo; 12–13 © Panoramic Images; 12a © John Nakata/Corbis; 12b and 14i © Bruce Amos/Shutterstock; 12c and 14g © Luis Castaneda Inc; 13a and 23 © PhotoLink; 13b © Rod Planck/Photo Researchers, Inc.; 13c and 14b © Ronald Sherwood/Shutterstock; 14a © Jakez/Shutterstock; 14c © Visual&Written SL/Alamy; 14d © Kenneth M. Highfill/Photo Researchers, Inc.; 14f © Rick & Nora Bowers/Alamy; 14h Rod Planck/Photo Researchers, Inc.; 15a © Supri Suharjoto/Shutterstock; 16 © Daniel Hebert/Shutterstock; 17b © Anthony Mercieca/Photo Researchers, Inc.; 18a © Joe McDonald/Corbis; 18c © Maximilian Weinzierl/Alamy; 18d © Index Stock Imagery, Inc.; 19 © Anna Galejeva/Shutterstock; 20 © Layne Kennedy/Corbis; 22a © Comstock Images/Alamy; 22b © Lee O'Dell/Shutterstock; 22c © Baloncici/Shutterstock; 24 © David Hughes/Shutterstock

Copyright © 2008 Millmark Education Corporation

Published by Millmark Education Corporation
PO Box 30239
Bethesda, MD 20824

ISBN-13: 978-1-4334-0002-5

Printed in the USA

10 9 8 7 6 5 4